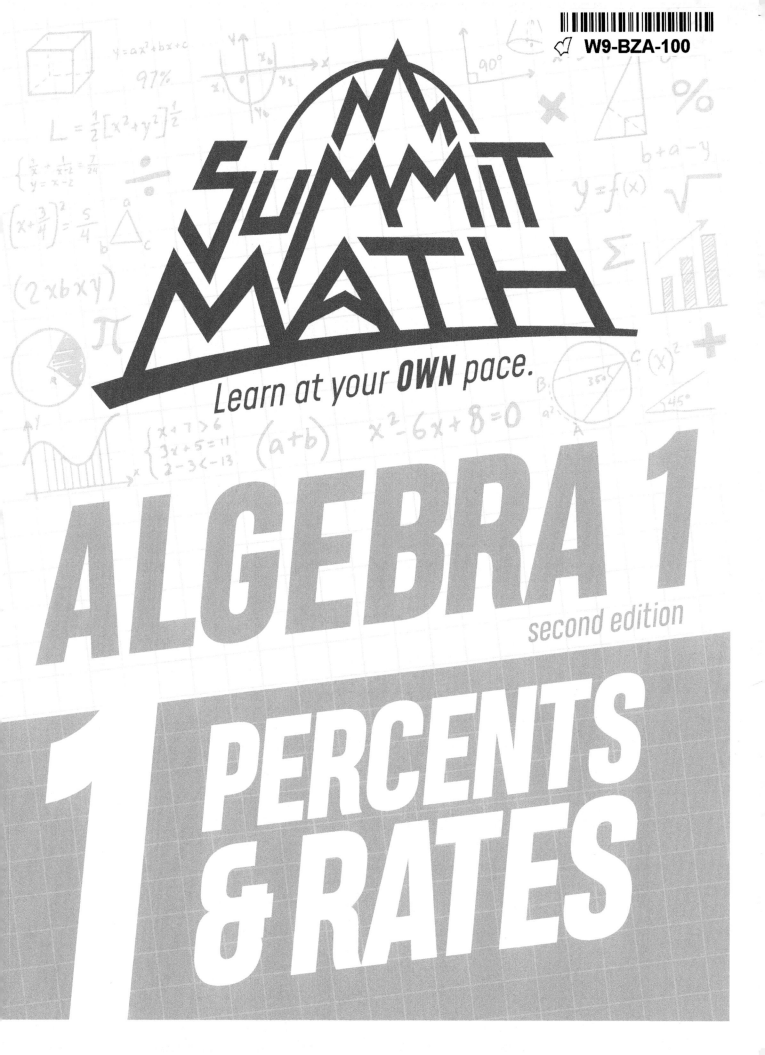

SUMMIT MATH

Learn at your **OWN** pace.

ALGEBRA 1

second edition

1 PERCENTS & RATES

DEDICATION
To Lauren, Chloe, Dawson and Teagan

ACKNOWLEDGEMENTS
I started writing these books in 2013 to help my students learn better. I kept writing them because I received encouraging feedback from students, parents and teachers. Thank you to all who have used these books, pointed out my mistakes, and made suggestions along the way. Thank you to all of the students and parents who asked me to keep writing more books. Thank you to my family for supporting me through every step of this journey.

This book was typeset in the following fonts:
Seravek + Mohave + *Heading Pro*

Graphics in Summit Math books are made using the following resources:
Microsoft Excel | Microsoft Word | Desmos | Geogebra | Adobe Illustrator

First printed in 2017

Printed in the U.S.A.

Summit Math Books are written by Alex Joujan.

www.summitmathbooks.com

INTRODUCTION

Learning math through Guided Discovery:
A Guided Discovery learning experience is designed to help you experience a feeling of discovery as you learn each new topic.

Why this curriculum series is named Summit Math:
Learning through Guided Discovery can be compared to climbing a mountain. Climbing and learning both require effort and persistence. In both activities, people naturally move at different paces, but they can reach the summit if they keep moving forward. Whether you race rapidly through these books or step slowly through each scenario, this curriculum is designed to keep advancing your learning until you reach the end of the book.

Guided Discovery Scenarios:
The Guided Discovery Scenarios in this book are written and arranged to show you that new math concepts are related to previous concepts you have already learned. Try to fully understand each scenario before moving on to the next one. To do this, try the scenario on your own first, check your answer when you finish, and then fix any mistakes, if needed. Making mistakes and struggling are essential parts of the learning process.

Homework and Extra Practice Scenarios:
After you complete the scenarios in each Guided Discovery section, you may think you know those topics well, but over time, you will forget what you have learned. Extra practice will help you develop better retention of each topic. Use the Homework and Extra Practice Scenarios to improve your understanding and to increase your ability to retain what you have learned.

The Answer Key:
The Answer Key is included to promote learning. When you finish a scenario, you can get immediate feedback. When the Answer Key is not enough to help you fully understand a scenario, you should try to get additional guidance from another student or a teacher.

Star symbols:
Scenarios marked with a star symbol ★ can be used to provide you with additional challenges. Star scenarios are like detours on a hiking trail. They take more time, but you may enjoy the experience. If you skip scenarios marked with a star, you will still learn the core concepts of the book.

To learn more about Summit Math and to see more resources:
Visit www.summitmathbooks.com.

GUIDED DISCOVERY SCENARIOS

As you complete scenarios in this part of the book, follow the steps below.

Step 1: Try the scenario.
Read through the scenario on your own or with other classmates. Examine the information carefully. Try to use what you already know to complete the scenario. Be willing to struggle.

Step 2: Check the Answer Key.
When you look at the Answer Key, it will help you see if you fully understand the math concepts involved in that scenario. It may teach you something new. It may show you that you need guidance from someone else.

Step 3: Fix your mistakes, if needed.
If there is something in the scenario that you do not fully understand, do something to help you understand it better. Go back through your work and try to find and fix your errors. Mistakes provide an opportunity to learn. If you need extra guidance, get help from another student or a teacher.

After Step 3, go to the next scenario and repeat this 3-step cycle.

NEED EXTRA HELP?
watch videos online

Teaching videos for every scenario in the Guided Discovery section of this book are available at www.summitmathbooks.com/algebra-1-videos.

CONTENTS

Section 1
INTRODUCTION TO PERCENTS

A percentage is one way to express a fraction. If you set a goal to read 10 pages in a book, after you have read 5 of them, you have finished 5 out of 10, or five-tenths of your goal. Using percentage language, you can say you have read 50 percent (50%) of your goal.

1. To learn the topics in this book, you should already be familiar with some percentage concepts. For example, you should know how to write a fraction as a percent.

    Without a calculator, how can you convert $\dfrac{1}{5}$ to a percent?

2. When it is difficult to convert a fraction to a percent, it is likely a better use of your time to use a calculator to do the conversion.

    Using a calculator, how can you convert $\dfrac{19}{40}$ to a percent?

3. What portion of the figure is shaded in each image below? Express this as a simplified fraction.

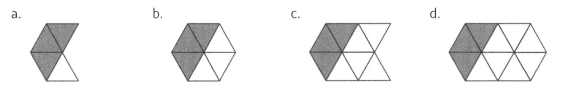

4. Convert each of the previous fractions to percents.

5. When you calculate a percent of another number, like 10% of 200, you identify a specific fractional amount of that number. Some percents are easier to compute. For example, what is 50% of 10?

4

6. Some percentages are more challenging but can still be computed without a calculator if you are familiar with common percents.

    a. How much money is 25% of $4?

    b. You bought 20 pounds of apples at an orchard and you used 75% of the apples to make applesauce. How many pounds of apples did you have left over?

Other percent calculations, like 35% of 43, or 12% of 112, require more time and can be done more quickly with a calculator. This brings up an important point, though. A calculator is only helpful if you input correct values.

7. For example, if you try to determine 50% of 10 by making a calculator compute 50x10, the calculator will give you a result of _500_. Explain why this result is not reasonable.

8. If you can estimate percentages before you do calculations, you might notice your errors if you make mistakes. Try to estimate each value below. Do <u>not</u> use a calculator.

    a. 48% of 12   b. 26% of 16   c. 99.1% of 20

A percent can also be more than 100%. For example, 300% of 20 means 3 sets of "100% of 20." Since 100% of 20 is exactly 20, then 300% of 20 is 3 sets of 20, which is 60.

9. Use a calculator to compute 364% of 45.

10. Sometimes it will be useful to write an expression like "22% of a number" in a smaller format, such as 22% of N, or even smaller as $0.22N$. The expression $0.22N$ shows that you can compute 22% of N by multiplying the decimal form of the percent by the variable N. Write an expression that represents each of the following, using decimals instead of percents.

    a. 50% of A b. 3% of B c. 275% of C d. 0.4% of D

Now that you have reviewed some percentage topics, this book will build on these topics and guide you through a selection of scenarios that use percents to make comparisons and calculations.

NOTES

Use this page to record important ideas in the previous section or
for any other writing that helps you learn the topics in this book.

Section 2

PERCENT CHANGES: INCREASES & DECREASES

11. A puffer fish has a circumference of 10 inches. When it feels threatened, it sucks in water and increases its circumference to 15 inches. To show a numerical change like this, you can compare the amount of the increase to the original amount. The fish grows by 5 inches, which is __50__ % of its original circumference. Using percents, the circumference grows by __50__ %.

12. If the fish continues to feel threatened, it expands its body more, until its circumference is 20 inches. Since the fish has now expanded by a total of 10 inches, it has increased its <u>original</u> circumference by __100__ %. This shows an interesting detail. If a number increases to become twice as large as its original value, it has increased by __100__ %.

13. If a number becomes three times as large as its initial value, by what percent has it increased?

14. If a number becomes one fourth of its original value, by what percent has it decreased?

15. Suppose that you measure your height one year and find that you are 50 inches tall. One year later, you measure again to check your growth and you find that your height has increased to 54 inches. In order to describe your growth as a percent increase you need to first consider your <u>amount</u> of growth, which is 4 inches.

    a. The <u>amount</u> of growth is a fractional amount of your original height, in this case $\frac{4}{50}$. This fraction can be simplified and written as the fraction __2/25__.

    b. Every fraction can be represented in decimal form. Use a calculator to write the value of $\frac{2}{25}$ in its decimal form.

    c. If you express 0.08 as a percent, it is __8__ %, which leads us back to the height scenario above. If your height increases from 50 inches to 54 inches, you can say that your height has increased by __8__ %.

16. Your height increases from 54 inches to 58 inches during the next year, another yearly increase of 4 inches. Does this represent an 8% increase again? If not, by what percent did your height increase if it changed from 54 to 58 inches?

The previous scenario involves challenging numbers. Return to simpler numbers for a moment in order to gain more familiarity with percentage changes.

17. A shirt initially costs $50, but it goes on sale for $40. Its price was decreased by _____%.

18. There are 80 students who are absent on Monday, and 100 students are absent on Tuesday. From Monday to Tuesday, the number of absences increased by _____%.

19. An $80 pair of boots decreases by $20 to sell for $60. A pair of shoes is priced at $60, but it only sells for $40. Which item had its price reduced by a greater percent, the boots or the shoes?

20. If 40 people earned a passing score last year, and only 20 people earned a passing score this year, by what percent did the number of passing scores decrease?

21. If you owe $20 to your friend today, and tomorrow you pay the entire amount that you owe, by what percent did you reduce the amount of money that you owed your friend?

22. When 100 increases to become 140, it might be easy for you to see that 100 has increased by 40 percent, because 100 is an easy number to work with. The number 100 is the foundation of percentages. The word percent comes from the phrase "per cent," which means "per 100." Try something more challenging to see if you understand the concept of percent change. If $140 increases to become $180, by what percent has it increased?

23. Fill in each blank below.

   a. If 100 increases by 20%, its new value is _____.

   b. If 100 decreases by 20%, its new value is _____.

   c. If 80 increases by 35%, its new value is _____.

   d. If 80 decreases by 35%, its new value is _____.

24. Write an expression that represents each of the following.

      a. *E* increases by 50%   b. *F* decreases by 25%

      c. 5% more than *G*   d. 15% less than *H*

25. In the previous scenario, you may not have realized that you can combine like terms. For example, an expression like *X* + 2*X* can be written as 3*X* and the expression 10*Y* − 3*Y* can be simplified to become 7*Y*. Look at your final expressions in the previous scenario and combine like terms.

26. Notice your expressions in the previous scenario. There are some common structures that you can identify in those expressions. Fill in each blank below.

      a. If *X* increases by 35%, its new value is _____.

      b. If *X* decreases by 19%, its new value is _____.

      c. If *X* increases by 100%, its new value is _____.

      d. If *X* decreases by 100%, its new value is _____.

27. Fill in each blank below.

      a. If *X* becomes 2.7X, it has increased by _____%.

      b. If *X* becomes 0.25X, it has decreased by _____%.

28. Compute the percent by which the initial value must change in order to become the final value.

      a. initial value: 100   final value: 120   percentage change: _____

      b. initial value: 100   final value: 90   percentage change: _____

      c. initial value: 100   final value: 20   percentage change: _____

29. Fill in the blank to show the percentage change.

    a. initial value: 100   final value: 220   percentage change: _____

    b. initial value: 100   final value: 300   percentage change: _____

    c. initial value: 100   final value: 450   percentage change: _____

30. Fill in the blank to show the percentage change.

    initial value: X final value: $0.28X$ percentage change: _____

31. Compare each expression shown below to an original amount of X. State whether X has increased or decreased and identify the percent by which X has changed.

    a. $1.64X$ b. $2.50X$ c. $0.99X$ d. $5X$

32. In 2013, through a generous anonymous donation, the high school was able to fund the construction of a larger seating area in their football stadium, increasing the seating capacity from 490 to 749. By what percent was the seating capacity increased?

33. In 2003, Elon Musk founded an electric car company named Tesla Motors. On September 9, 2012, the stock price of Tesla was $30 per share. On September 9, 2014, the stock price of Tesla was $283 per share. By what percent did the stock price increase from 2012 to 2014?

34. Between which two consecutive months did the population change by the greatest percent?

Month	April	May	June	July
Population	30,000	36,000	29,000	35,000

NOTES

Use this page to record important ideas in the previous section or
for any other writing that helps you learn the topics in this book.

Section 3

WRITING EQUATIONS TO CALCULATE PERCENTS

In this section, you will learn how to write equations to solve percent scenarios.

35. Write each statement as an equation, but do not solve the equation.

    a. The product of a number and 5 is 20.

    b. The quotient of 100 and a number is 50.

    c. 6 is 7 less than 3 times a number.

36. Write each statement as an equation, but do not solve the equation.

    a. A number is 3 more than 10.
    b. A number is 6 less than 4.

    c. 3 is 4 more than a number.
    d. A number increased by 5 is 15.

The previous two scenarios do not involve percentages, but they are intended to help you think about writing mathematical statements as equations. The next scenarios will focus only on percentages.

37. Write each statement as an equation. Do not solve the equation.

    a. 50% of a number is 45.

    b. A number is 24% of 80.

    c. 15 is a percentage of 20.

38. Consider the question, "What number is 45% of 18?" It is difficult to mentally compute this, but it is easier to write it out as a statement and convert that statement into an equation: "A number is 45% of 18." Write this statement as an equation.

     14

39. Read each question. Write an equation that matches the question. Do not solve the equation.

    a. What number is 12% of 30?   b. 12% of 30 is what number?

    c. 45 is 20% of what number?   d. 20% of what number is 45?

40. Answer each of the questions in the previous scenario by solving the equation you wrote.

41. Read each question and convert it into an equation. Do not solve the equation.

    a. 18 is what percent of 24?   b. What percent of 24 is 18?

    c. 36 is what percent of 15?   d. What percent of 15 is 36?

42. Answer each of the questions in the previous scenario by solving the equation you wrote.

43. Each of the following scenarios shows a percentage relationship that can be expressed as follows: ___% of ___ is ___. For each scenario, show the percentage relationship by writing an equation. Do <u>not</u> solve the equations.

     a. There are 200 animals that live at the zoo and 45% of the animals are birds. How many birds, *B*, live at the zoo?

$$200 \times .45$$

     b. Ilana made 65% of her free throws one day at practice. She made 70 free throws in all. How many free throws, *F*, did she attempt that day?

$$70 \times .65$$

     c. On average, there is precipitation on 122 days every year in New York City. What percent, *P*, of the 365 days in a year experience precipitation in New York City?

44. Answer each of the questions in the previous scenario.

45. Read each statement and convert it into an equation. Do not solve the equation.

     a. 50 is 25% more than a number.   b. *N* is 30 percent higher than 10.

     c. 32 is 20 percent below a number   d. 15% less than 20 is *N*.

46. Solve each of the equations you wrote in the previous scenario.

47. Read each statement and convert it into an equation. Then, solve the equation.

    a. After H decreases by 9%, it is equal to 50.

    b. The price of a gift is G. After a sales tax of 7% is added to its price, it costs a total of $80.

48. Read each scenario and write an equation that expresses the percentage relationship in that scenario. Do not solve the equations.

    a. The diagonal screen length of a phone increased by 20 percent. After the increase, it is 4.8 inches. What was the original diagonal screen length, L, of the phone?

    b. The average summer temperature in 2013 was 73.7°. This was 2% lower than the average summer temperature in 2012. What was the average temperature, T, in 2012?

49. Answer each of the questions in the previous scenario.

50. Read each scenario and write an equation that expresses the percentage relationship in that scenario. Do not solve the equations.

    a. A jacket's price tag is $200, but you will need more than that to buy the jacket because there is a sales tax of 8% that is added to the price. What is the total price of the jacket, *P*?

    b. You paid $23.85 for a meal, which included a 6% sales tax. What was the cost of the meal, C, before the tax was added?

51. Answer each of the questions in the previous scenario.

52. In 2013, Dee's house was worth 112% of what she paid when she bought the home three years earlier.

    a. By what percent did the value of Dee's house change during those three years?

    b. Lou's house is worth 85% of its purchase price five years ago. By what percent did the value of his house change during the last five years?

53. The price of a shirt is $24 on Friday, but the price increases 22% over the weekend. How much does that same shirt cost on Monday?

54. If the value of a car in June of 2012 was *V* dollars, but the value decreased 8% over the course of the next 12 months, what is the value of the car in June of 2013, written in terms of *V*?

55. On September 16, one share of Intel stock cost $21. That price was 5% higher than the previous day.

   a. Write an equation to express this price change, using the variable *p* to represent the price on September 15. Don't solve the equation.

   b. Make a guess: was the price on September 15th closer to $15, $20, or $25?

56. A percentage change can often be expressed with an equation. There are 3 common types of percentage change scenarios. Write an equation for each type of scenario below.

   a. Type 1: The price of a ball increases from $12 to $15. By what percent did its price change?

   b. Type 2: Before tax, the price of a ball is listed as $12. The sales tax is 7.4%. How much do you have to pay to buy the ball?

   c. Type 3: Your sales receipt shows that you paid $12 for a ball. If the sales tax was 7.4%, what was the original price of the ball as listed on its price tag?

57. The most challenging scenario is usually Type 3. Go back through every scenario in this section and mark each scenario as Type 1, Type 2, or Type 3.

NOTES

Use this page to record important ideas in the previous section or
for any other writing that helps you learn the topics in this book.

Section 4
VARIOUS PERCENT SCENARIOS

58. A park ranger measures a young redwood and finds that it is 20 feet tall. The following year, the tree has grown to a height of 22 feet. By what percent did the height increase?

59. Since a flight still has seats that have not been purchased, the price of a seat on the flight drops 25% to a new price of $300. What was the original price for the flight? To answer this question, think about which price changes by 25%.

    a. Does the original price change by 25% or does the new price change by 25%?

    b. What was the original price of the flight?

60. The football jerseys are not selling very quickly at a local store so the store manager lowers the price by 12% to $84.48. Sales immediately improve, which confirms that the price change was a good decision. What was the original price for a football jersey? Confirm that your answer is reasonable.

61. Mallory buys a new set of golf clubs and her receipt shows that she paid $845.30. If there is a sales tax of 7% included in her receipt, what was the cost of the golf clubs before taxes were calculated?

62. In 2009, the school's energy bill for the month of August was $33,862. After installing solar panels on the roof in 2012, the energy bill for the month of August was $19,036. Calculate the percent change in the school's August energy bill from 2009 to 2012.

63. Sarah's investment in Jet Blue stock grew 28% to become $448. How much did she originally invest?

64. ★Time Magazine published an article stating that some breakfast cereals are more than 50% sugar by weight and suggested that kids eat 10 pounds of sugar each year from breakfast cereal alone. Kellogg's Honey Smacks are 56% sugar by weight and a box of Honey Smacks weighs 15.3 ounces. Estimate the number of boxes of Honey Smacks that a child will need to consume in one year to ingest 10 pounds of sugar from that cereal alone.

65. Consider the following statement. "Of all the people earning around $60,000 per year, the median value of their retirement account is $105,000. This is 74% below what they should have saved when they retire." What assumption is this statement making about how much a person should have saved when they retire?

66. In 1991, the city of Boston began a highway project known as the Big Dig. When it was completed in 2006, the cost of the project totaled approximately $22 billion, which was 686% higher than the original estimate. How much did the city of Boston originally expect to pay for this project?

67. John's money manager called him and informed him that his investment of $18,000 had decreased by 108%. John was understandably very upset and as he thought more about it, he realized that something seemed peculiar about this scenario. Can you explain John's dilemma?

68. ★In 2015, it was estimated that the price of used cars would decrease by 2.5 percent every year for the next three years.

    a. As a result, used car prices should have dropped a total of ____ percent from 2015 to 2018.

    b. If this decline in prices continued for 7 more years after 2018, by what percent would used car prices decrease from 2015 to 2025? (Since the price decreased by 2.5 percent every year for 10 years, by what percent would you guess that the price decreased by the end of the 10 years? Notice how this compares with what actually happens.)

69. Why is it that a 50% decrease, followed by another 50% decrease, does not combine to produce an overall decrease of 100%?

70. Read the statement below, which originates from a report released in 2010.

    "55 percent of India's population of 1.1 billion, or 645 million people, are living in poverty."

    Although this finding is troubling, it is also inaccurate. Why is this statement false? How could you change it to make it accurate?

71. In a survey, each participant was asked one question, "If you could only own one pet, which pet would you choose?" Results from the survey are shown. Some participants selected an animal that is not listed in the graph so their responses are not visible.

    a. Estimate the total number of participants who answered the survey question.

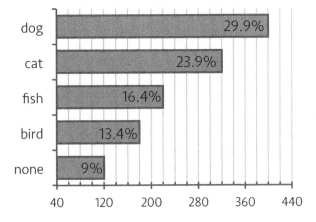

    b. What percent of the participants did not select one of the 5 categories shown?

24

72. ★An electronics store just opened and it is offering various sales to bring in customers. One sale features a cell phone with an original price of $650. The phone is on sale for 20% off the original price and the customer can pay for the phone gradually by making 12 equal monthly payments over the course of one year. A 6.5% tax is applied after the discount.

    a. How much would a customer pay each month for this phone?

    A second phone is offered for 30% off its original price. After the discount, the tax for this phone is $26.39.

    b. What is the total price of the second phone after the tax is applied?

    c. What was the original price of the second phone, not including tax?

You have worked through many percent scenarios so far, but the next scenario is quite complex. If it seems too challenging, go on to the scenarios that follow and come back to this one afterward.

73. ★Shake Shack introduces a new menu item, the Mintmallow Mud milkshake, but the initial price seems to be too high. In order to increase sales, Shake Shack lowers the price by 20%. After they realize that they can charge more and still maintain a high demand, they increase the price 10%, which brings the price to $3.90. What was the initial price of the Mintmallow Mud milkshake?

74. ★One aspect that makes the previous scenario confusing is that you are changing an unknown price by two different percentages and it is difficult to express this. Start with some specific numbers to build up your ability to handle unknown numbers.

    a. Start with 100. Increase 100 by 20%.

    b. Decrease your answer to part a. by 20%. What number do you end up with?

75. Start over with 100 again.

    a. Decrease 100 by 10%.

    b. Increase your answer to part c. by 10%. What number do you end up with?

76. When you decrease 100 by 10% and then increase the result by 10%, why don't you end up back where you started?

77. ★Start with another number. Instead of a specific number, generalize this and start with X.

    a. Increase X by 20%. What is the simplified expression that represents your result?

    b. Decrease your answer to part a. by 20%. What is your simplified result?

78. ★Start with H.

    a. Decrease H by 10%. What is your simplified result?

    b. Increase your answer to part a. by 10%. What is your simplified result?

79. ★Start with Z.

    a. Increase Z by 10%. What is your result?

    b. Decrease your answer to part a. by 20%. What is your result?

80. ★Go back to the milkshake scenario and see if you understand how to work through it now.

81. The chart shows the average temperature during the first 8 months of the year.

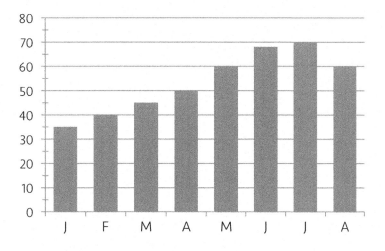

a. By what percent did the average temperature change from April to May?

b. By what percent is the highest temperature more than the lowest temperature?

c. By what percent is the lowest temperature less than the highest temperature?

82. Suppose that you work at a shoe store and your boss comes over to you with a new sales idea. She says, "Change all of the price tags in the store! Raise all of the prices in the store by 50% and then put large signs on the store windows that say, 'All items are 50% off of the listed price.' This way, the customers will feel like they are saving money, but they will actually still be paying the same price that we originally listed on all of our price tags. Great idea, right?" Do you think this suggestion will boost sales?

83. You have 3 coupons that can be used at your favorite shoe store. One coupon will reduce the price of an item by 10%, another is for a 25% discount, and the third is for a 40% discount. On this special occasion, the store will let you use all 3 of them, in any order that you want.

    a. If you buy a pair of shoes and use all of the coupons, what is largest total percentage discount that you can get?

    b. How much would you pay for a pair of $200 shoes? Assume there are no taxes.

84. ★You set a goal to finish the first trimester with an average of at least 86.5%. There are five assessments that will determine your final average. After the first four assessments, which are worth a total of 175 points, you have an 84% average. How many points do you need to earn on the final assessment if it is worth 65 points to achieve your goal?

85. ★The height of a tree increases by 10% every year for 5 years in a row. By what percent did the height of the tree increase over the entire course of that 5-year period?

86. The value of a car decreases by 10% every year. After 5 years, will the value of the car have decreased by more than, less than, or exactly 50%? How do you know this?

NOTES

Use this page to record important ideas in the previous section or for any other writing that helps you learn the topics in this book.

Section 5
CUMULATIVE REVIEW: PART 1

87. Simplify each expression shown.

    a. $12.5 - 10.45$ b. $\dfrac{5}{12} \cdot \dfrac{9}{10}$ c. $5 \div \dfrac{5}{11}$

88. Which of the four fractions does not represent the same percentage value as the other three?

    a. $\dfrac{-3}{6}$ b. $\dfrac{2}{-4}$ c. $-\dfrac{5}{10}$ d. $\dfrac{-1}{-2}$

89. A circle is placed on top of a rectangle, as shown. Which shape has a larger area?

10 cm

8 cm

90. Near the end of the season, two baseball teams had each played a total of 143 games. The table displays how many runs each team had scored at that point in the season, as well as the number of runs scored by their opponents. Using the data, which team had won more of their games? Explain your reasoning.

Team	Runs Scored by Team	Runs Scored by Opponents
Colorado Rockies	757	740
Kansas City Royals	591	610

91. In the previous scenario, the Rockies had won 68 games and lost 75 games at that point in the season. The Royals had won 75 games and lost 68 games. How could that have happened?

92. Which value is larger?

    Option #1: 50% of 10% of $100   Option #2: 10% of 50% of $100?

NOTES

Use this page to record important ideas in the previous section or
for any other writing that helps you learn the topics in this book.

32

Section 6
INTRODUCTION TO RATES

Consider scenarios that involve a rate, which is a ratio that identifies how a change in one measurement is related to a change in another measurement. You use rates more often than you may even realize.

93. If you can buy 12 bottles of water for $6, how much do you pay per bottle?

94. Avik slept for a total of 58 hours during one full week. If he got about the same amount of sleep each night, estimate the number of hours that he slept each night that week.

95. Aimee buys a lot of candy. She spends an average of $12 every 2 weeks. How much will she spend on candy during the next 15 weeks?

96. Jon rides his bike 30 miles in 2.4 hours. Kyle rides his bike 18 miles in 1.5 hours.

   a. Who travels at a faster rate?

   b. How far would each rider travel in 5 hours?

97. For the team picnic next week, everyone brings a different food item. You are asked to bring ice. At the grocery store, ice comes in two bag sizes. There are 8-pound bags of ice that cost $2.20 and there are 20-pound bags that cost $4.25. If you need to buy a total of 80 pounds of ice, how much money can you save if you only buy the bags that are a better deal?

98. The Jain family likes to pick blueberries. The blueberry farm that they go to is very popular so the farm charges an entrance fee of $10. One weekend, they picked 6 pounds of blueberries and paid a total of $40. On another weekend, they paid a total of $70 for 12 pounds of blueberries. How much does the farm charge for the blueberries?

99. ★John and Anna agree to sort cans at a local Food Bank on Saturday morning. John gets there first, finds a pile of 650 cans, and begins sorting them into boxes at a rate of 10 cans per minute. Anna arrives five minutes later and starts sorting cans at a rate of 14 cans per minute. After they have sorted all of the cans, Anna claims that she did more work than John. How many cans did Anna sort?

100. For a service project, you devote several days to rebuilding a rock wall along a bike path. When you stopped working yesterday, you had rebuilt a total of 22 feet of the wall. After working 4 more hours today, you have rebuilt a total of 28 feet of the wall.

    a. If you work at a constant rate, how many feet do you rebuild each hour?

    b. How many hours will you take to rebuild the entire wall if it is 48 feet long?

101. What is the rate that is described in the next scenario?

102. ★Liz runs a sports camp and she uses a machine to paint all of the lines on the fields during the day before the camp opens. There are 12,960 feet of lines that need to be painted. When she takes a lunch break at noon, she has painted 4,500 feet of lines. If she paints at a rate of 90 feet per minute, how long will it take her to paint the rest of the lines?

103. ★A passenger train leaves a station at noon to travel nonstop to Boston. At 12:30pm, the train is 288 miles from Boston and at 4:00pm, the train is 120 miles from Boston.

    a. How far is the station from the Boston?

    b. At what time will the train reach Boston?

104. At the end of the season in 1971, Coach Miller Bugliari had 116 wins as a soccer coach. At the end of the season in 2013, he had 785 wins.

    a. Calculate the percent change in his wins from 1971 to 2013.

    b. At the end of the season in 2015, Coach Bugliari had 820 wins. Did Coach Bugliari win games at a faster rate between 1971 and 2013 or between 2013 and 2015?

NOTES

Use this page to record important ideas in the previous section or for any other writing that helps you learn the topics in this book.

Section 7
USING GRAPHS TO CALCULATE RATES

105. You have two pet fish, and you have named them Linus and Lucy. Your fish food is in the form of little pellets. As your fish grow, the number of pellets that you need to feed them increases. On the pellet box, the directions state that you should give your fish 1 pellet per day plus an additional 2 pellets per centimeter, based on their length.

a. If Linus is 4 cm long and Lucy is 5 cm long, how many <u>more</u> pellets should you give Lucy each day?

b. In the graph, show the amount of food you should give to your pets as they grow.

c. Your neighbor Jamey has a fish named Eustace that gets 17 pellets per day. How long is Eustace?

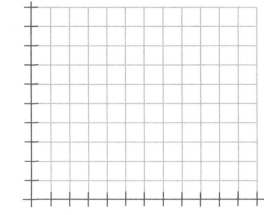

Fish length, in cm

106. The Jain family goes to a local farm and picks strawberries. The cost of the strawberries depends on the amount of strawberries they pick.

a. What is the rate shown in the graph? Identify the numerical value of the rate and express the rate using proper units.

b. Write an equation that shows the total cost, C, if the family picks *p* pounds of strawberries.

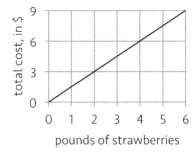

107. A small earthworm crawls at a rate of 0.25 centimeters per second.

a. How long will it take for the worm to travel 3 centimeters?

b. How long will it take for the worm to travel 3 meters?

108. A larger earthworm crawls at a rate of 0.5 centimeters per second.

a. In the graph to the right, show how the distance of the larger earthworm increases each second.

b. In the same graph, show how the distance of the small earthworm (in the previous scenario) increases each second. Use a dashed line to show the small earthworm and a solid line to show the larger earthworm.

c. If the earthworms raced each other, how far ahead would the larger earthworm be after 12 seconds?

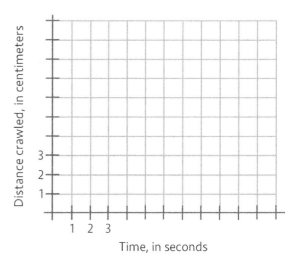

109. Christine and Kerri are sisters and they both work as lifeguards at the same pool, although one of them earns more because she has worked there for three years longer than her sister.

a. Which sister started working at the pool first?

b. How much does each sister earn per hour?

Kerri

hours worked	money earned
2	$26
5	$65
11	$143

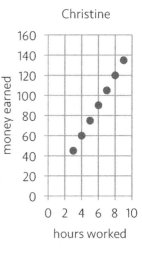

Christine

c. Last week, they both worked the same number of hours, but Christine earned $20 more than Kerri. How many hours did they each work last week?

110. For his birthday party, Nolan wants to invite his friends to rent a sailboat for a ride around a nearby lake. The cost of the sailboat rental depends on the number of people who will attend the party.

a. How much would it cost for 8 people to attend?

b. Write an equation that shows the total cost of the sailboat rental, C, if n people attend the party.

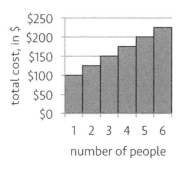

111. Clyde's family reunion is coming up and each person has a different job to do to prepare the food for the reunion picnic. Clyde's job is to cook a large amount of chicken. He does some research to find out how much chicken he should cook and he finds the graph shown to the right.

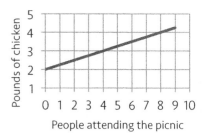

a. What rate can be determined from the graph?

b. Before Clyde started to cook the chicken, he found out that 1 more person was coming. How much more chicken will he prepare for that extra person?

c. How much chicken does Clyde need to cook if 14 people are going to attend the picnic?

d. Last year, Clyde cooked 8 lbs. of chicken. How many people were at the picnic last year?

112. The graph displays how the weight of a typical candle changes as it burns.

a. At what rate does the weight of the candle decrease, in ounces per hour?

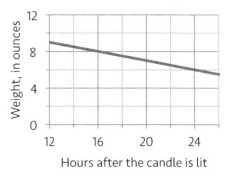

b. At what rate does the weight of the candle decrease, in ounces per day?

★c. How heavy was the candle when it was lit?

NOTES

Use this page to record important ideas in the previous section or for any other writing that helps you learn the topics in this book.

Section 8
RATES IN EQUATIONS

113. Consider the equation $y = -5x$.

    a. What is y if $x = 2$?

    b. What is x when $y = 50$?

114. Consider the equation $N = 6d - 15$.

    a. Find the value of N when d is 4.

    b. What is the value of d when $N = -45$?

115. Harriet works for her uncle one summer and at the end of the summer, her payment is calculated by using the equation $P = 120d$, where d is the total number of days that she works for her uncle and P is the total value of her paycheck, measured in dollars.

    a. What does the 120 represent in this equation?

    b. How much is Harriet paid if she works for her uncle for 5 days?

116. The movement of a car is represented by the equation $d = 40h$, where d is the distance the car has traveled, in miles, after h hours.

    a. What does the 40 represent in this equation?

    b. How far has the car traveled after 5 hours?

117. If a gift card has not been used for several years, the gift card company may charge a fee for each month of inactivity. For example, after three years of inactivity, suppose the value of a gift card is represented by the equation $V = 150 - 2m$, where V is the value of the gift card, in dollars, after the card has been inactive for m months beyond the three-year limit.

    a. In this equation, what do the numbers 150 and 2 represent?

    b. What is the value of the gift card after it has been inactive for 1 year?

118. A restaurant has two types of tables, square ones and rectangular ones. The number of people that can be seated at the restaurant is given by the equation $4s + 6r = 92$.

    a. What do the s and the r represent in the equation?

    b. What do the 4 and the 6 represent in the equation?

    c. How many people can be seated at the restaurant?

119. Finn rides the bus to visit a friend in another state. On the highway, the bus maintains a speed of 60 miles per hour.

    a. What is the rate described in this scenario?

    b. After 3 hours, how many miles has Finn traveled?

    c. After H hours, how many miles has Finn traveled?

    d. If Finn has traveled exactly 200 miles at the moment the clock displays that the time is 1:00pm, what time will the clock display at the moment he has traveled exactly 500 miles?

120. When a valve is opened, water flows through a pipe at a rate of 30 gallons per minute.

    a. If the water is flowing into a tank that already contains 210 gallons, write an expression for the amount of water in the tank after the valve has been open for m minutes.

    b. Use your expression in part a. to determine what time the tank will be completely filled with water if the tank can hold 450 gallons and the valve is opened at 3:00pm.

121. An airplane lifts off the runway and gradually rises into the sky. After it reaches its cruising altitude, the airplane maintains a constant rate. Its motion is modeled by the equation $D = 5,400 - 640H$, where D is the distance, in miles, between the plane and its final destination and H is the number of hours after 3pm, when the plane begins cruising at a constant rate.

    a. How far is the airplane from its final destination at 3pm?

    b. At what time is the airplane 1,880 miles away from its destination?

    c. At what rate is the airplane flying, in miles per hour?

122. A computer transfers a file at a rate of 500 kilobytes per second.

    a. How long will it take to transfer the entire file if its original file size is 30 megabytes? A megabyte is approximately 1000 kilobytes.

    b. Write an equation that shows how many megabytes, M, still need to be transferred after t seconds have passed.

123. While downloading a very large file, you notice that the file is being transferred at a rate of 1.5 megabytes (MB) per second. At 8:00am, you notice that you have downloaded 145 MB. What time will you finish downloading the entire file if the total file size is 1000 MB (1 gigabyte)?

124. Abigail has been measuring the height of a bamboo plant for a science experiment. She sees that the height of the bamboo increases by the same amount every day. She models the growth of the plant with the equation $H = 3.5d + 4.9$, where H is the height in inches and d is the number of days that have passed since April 1, when she started recording measurements. For example, on April 4, $d = 3$, since 3 days have passed since April 1.

    a. What is the value of d on April 10 and what is the height of the bamboo plant on April 10?

    b. What is the height of the bamboo plant on April 2?

    c. What is the height of the bamboo plant on April 1?

    d. On what day will the height of the bamboo plant reach 25 inches?

125. In the previous scenario, what is the rate at which the plant is growing, in inches per day?

126. When Paulo's pool is full, it holds 16,000 gallons. After a leak allowed the pool to lose most of its water, it now contains only 3,400 gallons. After the leak is repaired, a pipe is turned on to begin filling the pool again. Water flows through the pipe at a rate of 45 gallons per minute.

    a. How much water is in the pool after 10 minutes?

    b. How much water is in the pool after m minutes?

    c. How many minutes will it take to fill the pool?

    ★d. If the pipe is turned on at 1:00pm, at what time will the pool be completely filled?

127. In 2012, the New Jersey Turnpike collected $992 million in revenue. This was the highest revenue of any toll road in the nation. When drivers use the NJ Turnpike, they pay an initial fee of $0.50 plus an additional 11 cents per mile driven.

    a. How much would you pay if you drove 1 mile on the NJ turnpike?

    b. How much would you pay to drive m miles on the turnpike?

    c. How many miles, m, can you drive on the turnpike if you want to keep your toll payment under $10?

128. After saving nickels, dimes and quarters for several years, Shauna now has $62.85 in her piggybank. The amount of money in Shauna's piggybank can be represented by either equation below.

    Equation #1: $\underline{\quad}n + 0.10d + 0.25q = 62.85$ Equation #2: $5n + 10d + 25q = \underline{\quad}$

    a. What do the letters n, d, and q represent in the two equations?

    b. What does 0.10 represent in Equation #1?

    c. What does 5 represent in Equation #2?

    d. Fill in the blanks to complete each equation above.

NOTES

Use this page to record important ideas in the previous section or
for any other writing that helps you learn the topics in this book.

48

CUMULATIVE REVIEW: PART 2

129. Identify the pattern in each table below and use this pattern to fill in the missing cell.

a.

x	y
−6	−1
−2	1
2	3
4	4
	12

b.

x	y
−2	7
0	3
2	−1
3	−3
6	

130. Plot the ordered pairs shown in the two tables above. Only plot points that fit in the graph. One point is done for you.

a.

b.

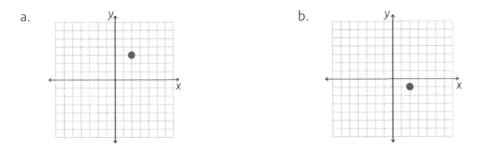

131. The sales tax in Washington, D.C. is 10%. If a meal costs $23.98 after the tax is calculated, how much did the meal cost before the tax was included?

132. In 2013, an estimated 32,000,000 tons of plastic waste was produced in the U.S.

   a. If the population of the U.S. in 2013 was 318,000,000 people, how many pounds of plastic waste did each person produce that year, on average?

   b. In 2013, how much plastic waste was produced per person each week?

133. Suppose 12 school buses transport 648 students to school and 160 cars transport 240 students to school. If all of the school buses were removed from service, how many extra cars would be needed to transport all of the remaining students?

NOTES

Use this page to record important ideas in the previous section or
for any other writing that helps you learn the topics in this book.

Section 10
ANSWER KEY

1.	Other options exist, but one way is to change the denominator to 100. This makes it easy to state the fraction as a percent. $\dfrac{1}{5} \cdot \dfrac{20}{20} \rightarrow \dfrac{20}{100}$ "20 hundredths" is 20 per 100 which has the same meaning as 20%.
2.	Find the decimal form of the fraction. In the calculator, divide 19 by 40. To do this, input $19 \div 40$ or $19/40$. As a decimal, $19/40$ is 0.475, which is 475 thousandths or 47.5 hundredths. 47.5 per 100 is 47.5 percent (47.5%). You may have learned to move the decimal 2 places to the right to convert a decimal to a percent, and this is true but it is important to know why that works.
3.	a. $\dfrac{3}{4}$ b. $\dfrac{3}{6} \rightarrow \dfrac{1}{2}$ c. $\dfrac{3}{8}$ d. $\dfrac{3}{10}$
4.	a. 75% b. 50% c. 37.5% d. 30%
5.	One-half of 10, which is 5.
6.	a. \$1 b. 5 pounds (75% of 20 is 15)
7.	500; 50% of a number is one-half of that number. One-half of 10 is just 5, so a result of 500 is not reasonable.
8.	a. ≈6 (about 50% of 12) b. ≈4 (about 25% of 16) c. ≈20 (about 100% of 20)
9.	$3.64 \times 45 \rightarrow 163.8$
10.	a. $0.5A$ b. $0.03B$ c. $2.75C$ d. $0.004D$
11.	50%
12.	100%
13.	200%
14.	75%
15.	a. $\dfrac{2}{25}$ b. 0.08 c. 8
16.	Approx. 7.4% $\left(\dfrac{4}{54} \text{ is } 0.0\overline{74} \text{ in decimal form} \right)$
17.	20%
18.	25%
19.	the shoes $\left(33\dfrac{1}{3}\% \text{ vs. } 25\% \right)$

20.	50%
21.	100%
22.	$\dfrac{40}{140} \rightarrow 0.286 \rightarrow$ approx. 28.6%
23.	a. 120 b. 80 c. 108 d. 52
24.	a. $E + 0.5E$ b. $F - 0.25F$ c. $G + 0.05G$ d. $H - 0.15H$
25.	a. $1.5E$ b. $0.75F$ c. $1.05G$ d. $0.85H$
26.	a. $1.35X$ b. $0.81X$ c. $2X$ d. 0
27.	a. 170 b. 75
28.	a. 20% (increase) b. 10% (decrease) c. 80% (decrease)
29.	a. 120% (increase) b. 200% (increase) c. 350% (increase)
30.	72% (decrease)
31.	a. increase 64% b. increase 150% c. decrease 1% d. increase 400%
32.	$\dfrac{259}{490} \rightarrow 0.529 \rightarrow$ approx. 52.9%
33.	$\dfrac{253}{30} \rightarrow 8.4\overline{3} \rightarrow 843\dfrac{1}{3}\%$
34.	Between June and July (20.7% increase)
35.	a. $5n = 20$ b. $100 \div n = 50$ c. $6 = 3n - 7$
36.	a. $n = 10 + 3$ b. $n = 4 - 6$ c. $3 = n + 4$ d. $n + 5 = 15$
37.	a. $0.5n = 45$ b. $n = 0.24(80)$ c. $15 = \dfrac{n}{100}(20)$ or $15 = n(20) \rightarrow$ if n is the decimal form of the percentage.
38.	$n = 0.45(18)$
39.	a. $X = 0.12(30)$ b. $0.12(30) = X$ c. $45 = 0.2X$ d. $0.2X = 45$
40.	a,b. 3.6 c,d. 225
41.	a. $18 = \dfrac{x}{100}(24)$ b. $\dfrac{x}{100}(24) = 18$ c. $36 = \dfrac{x}{100}(15)$ d. $\dfrac{x}{100}(15) = 36$
42.	a, b. 75% c, d. 240%
43.	a. 45% of 200 is $B \rightarrow 0.45(200) = B$ b. 65% of F is 70 $\rightarrow 0.65F = 70$

	c. P% of 365 is 122 $\rightarrow \dfrac{P}{100}\cdot 365 = 122$
44.	a. 90 birds b. ≈108 attempts c. ≈33.4%
45.	a. $50 = N + .25N \rightarrow 50 = 1.25N$ b. $N = 10 + .3(10) \rightarrow N = 1.3(10)$ c. $32 = N - .2N \rightarrow 32 = 0.8N$ d. $20 - .15(20) = N \rightarrow (0.85)20 = N$
46.	a. 40 b. 13 c. 40 d. 17
47.	a. $H - 0.09H = 50 \rightarrow 0.91H = 50 \rightarrow H \approx 54.9$ b. $G + 0.07G = 80 \rightarrow 1.07G = 80 \rightarrow G \approx \$74.77$
48.	a. $L + 0.2L = 4.8 \rightarrow 1.2L = 4.8$ b. $T - 0.02T = 73.7 \rightarrow 0.98T = 73.7$
49.	a. 4 inches b. ≈75.2°
50.	a. $200 + 0.08(200) = P \rightarrow 1.08(200) = P$ b. $C + 0.06C = 23.85 \rightarrow 1.06C = 23.85$
51.	a. $216 b. $22.50
52.	a. Increase 12% b. Decrease 15%
53.	$24 + .22(24)$ or $1.22(24)$ or $29.28
54.	$V - .08V$ or $0.92V$
55.	a. $p + .05p = 21$ or $1.05p = 21$ b. $20
56.	a. $12 + 12x = 15$ (x is the percent) b. $12 + .074(12) = x$ (x is the amount you pay) c. $x + .074x = 12$ (x is the original price)
57.	Type 1: None in this section Type 2: 41b, 41d, 46a, 49 Type 3: 41a, 41c, 43a, 43b, 44a, 44b, 46b, 51a
58.	10%
59.	a. original price b. $400 (solve .75x=300)
60.	$96
61.	$790
62.	Approx. 43.8% decrease
63.	$350
64.	around 20 boxes (≈ 18.7 boxes is more exact)
65.	This statement assumes that a person earning around $60,000/year should have about $403,846 saved when they retire.
66.	$2.8 billion
67.	If it lost all of its value, it would have decreased 100%. Once it has lost all of its value, it cannot decrease any more, so a 108% decrease is impossible.
68.	a. ≈7.31% b. ≈22.4%
69.	the two 50% decreases are 50% of different amounts
70.	55% of 1.1 bil. is only 605 mil. 645 mil. is approx. 58.6% of 1.1 bil.
71.	a. ≈1340 b. 100/1340 ≈ 7.5%
72.	a. price: $520; tax: $33.80; payment: $46.15 b. price: $406 + $26.39 = $432.39 c. $580
73.	Approx. $4.43

74.	a. $100+0.2(100) = 120$ b. $120-0.2(120) = 96$
75.	a. $100 - 0.1(100) = 90$ b. $90 + 0.1(90) = 99$
76.	After you increase 100 by 10%, it is bigger than 100. When you decrease the result by 10%, the amount of decrease is different than the amount of increase.
77.	a. $X + 0.2X = 1.2X$ b. $1.2X - 0.2(1.2X) = 0.96X$
78.	a. $H - 0.1H = 0.9H$ b. $0.9H + 0.1(0.9H) = 0.99H$
79.	a. $Z + 0.1Z = 1.1Z$ b. $1.1Z - 0.2(1.1Z) = 0.88Z$
80.	–
81.	a. 20% b. 100% more c. 50% less
82.	A 50% increase followed by a 50% decrease makes the new prices lower than the original prices, which will be good for customers but it might be bad for the store.
83.	a. 59.5% b. $81
84.	At least 61 out of 65 (≥ 60.6)
85.	Approximately 61%
86.	Less than 50%. After the first 50% decrease, the value is smaller, so the second 50% is a smaller dollar amount as well.
87.	a. 2.05 b. $\dfrac{3}{8}$ c. 11
88.	$\dfrac{-1}{-2}$ (option d.)
89.	The rectangle (80 cm^2) is larger than the circle (≈78.5 cm^2).
90.	There is not enough information to know how many games were won, but Colorado likely won more games because they scored more total runs than their opponents.
91.	Colorado lost many games by a small number of runs and won many games by a large number of runs.
92.	They are both the same value ($5).
93.	$6÷12 \rightarrow$ 50 cents per bottle
94.	58 hours÷7 → about 8.3 hours per night
95.	$6 per week → $6×15 = $90
96.	a. Jon (12.5 mph vs 12 mph) b. Jon: 62.5 miles; Kyle: 60 miles
97.	You will save $5. You spend $22 if you buy ten 8-lb. bags, but you only spend $17 if you buy four 20-lb. bags.
98.	$5 per pound. For either weekend, subtract $10 to get the amount paid for blueberries. Divide by the number of pounds to get the cost per pound.

99.	John sorts 50 cans by himself. Together, they sort 24 cans per min. They sort the remaining 600 cans in 25 minutes. Anna sorts 350 cans in 25 minutes.
100.	a. 6 feet every 4 hours, or 1.5 feet per hour b. 32 total hours
101.	90 feet per minute
102.	94 minutes solve: $4,500 + 90m = 12,960 \rightarrow m = 94$
103.	The train travels 168 miles in 3.5 hours, or 48 miles per hour. a. 312 miles b. 6:30pm
104.	a. Approx. 577% b. Faster rate from 2013-2015 → 15.9 wins/yr from 1973-2013 and 17.5 wins/yr from 2013-2015
105.	a. 2 more pellets 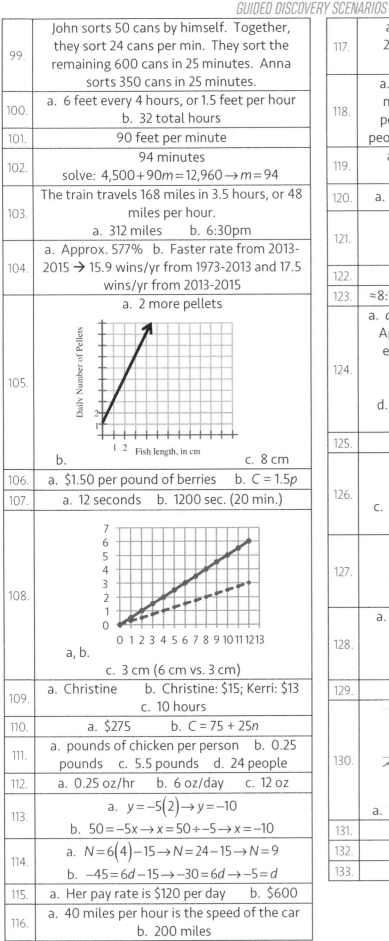 b. c. 8 cm
106.	a. $1.50 per pound of berries b. $C = 1.5p$
107.	a. 12 seconds b. 1200 sec. (20 min.)
108.	a, b. c. 3 cm (6 cm vs. 3 cm)
109.	a. Christine b. Christine: $15; Kerri: $13 c. 10 hours
110.	a. $275 b. $C = 75 + 25n$
111.	a. pounds of chicken per person b. 0.25 pounds c. 5.5 pounds d. 24 people
112.	a. 0.25 oz/hr b. 6 oz/day c. 12 oz
113.	a. $y = -5(2) \rightarrow y = -10$ b. $50 = -5x \rightarrow x = 50 \div -5 \rightarrow x = -10$
114.	a. $N = 6(4) - 15 \rightarrow N = 24 - 15 \rightarrow N = 9$ b. $-45 = 6d - 15 \rightarrow -30 = 6d \rightarrow -5 = d$
115.	a. Her pay rate is $120 per day b. $600
116.	a. 40 miles per hour is the speed of the car b. 200 miles

117.	a. 150: initial value of gift card is $150 2: the fee for inactivity is $2 per month b. $V = 150 - 2(12) \rightarrow 150 - 24 \rightarrow \$126$
118.	a. s: the number of square tables; r: the number of rectangular tables b. 4: 4 people can sit at each square table; 6: 6 people per rectangular table c. 92 people
119.	a. 60 miles per hour b. 180 mi. c. $60H$ mi d. 6:00 pm
120.	a. $210 + 30m$ b. About 3:08pm $(m = 8)$
121.	a. At 3pm, $H = 0$, so $D = 5,400$ b. When $D = 1,880$, $H = 5.5$, which represents 8:30pm. c. 640 mph
122.	a. 60 sec b. $M = 30 - 0.5t$
123.	≈8:10am (solve $145 + 90m = 1000 \rightarrow m = 9.5$)
124.	a. $d = 9$ on April 10, because it is 9 days after April 1. To find the height, let $d = 9$ in the equation → $H = 3.5(9) + 4.9 \rightarrow H = 36.4$ b. When $d = 1$, $H = 8.4$ in. c. Let $d = 0 \rightarrow H = 4.9$ in. d. Let $H = 25$ and solve for d. $d \approx 5.7 \rightarrow d \approx 6$ → approximately April 7th
125.	3.5 inches per day
126.	a. $3,400 + 45(10) \rightarrow 3,850$ gallons b. $3400 + 45m$ c. solve: $3,400 + 45m = 16,000 \rightarrow 280$ min. d. 5:40pm
127.	a. 61 cents ($0.61) b. $0.50 + 0.11m$ or $50 + 11m$ c. solve: $0.50 + 0.11m < 10 \rightarrow m < 86.4$ At most 86 miles
128.	a. the number of nickels, dimes, quarters b. the value of a dime, in dollars c. the value of a nickel, in cents d. Eq.#1: 0.05 Eq.#2: 6,285
129.	a. $x = 20$ b. $y = -9$
130.	a. b.
131.	$21.80
132.	a. ≈201.3 lbs. b. ≈3.87 lbs.
133.	432 cars

HOMEWORK & EXTRA PRACTICE SCENARIOS

As you complete scenarios in this part of the book, you will practice what you learned in the guided discovery sections. You will develop a greater proficiency with the vocabulary, symbols and concepts presented in this book. Practice will improve your ability to retain these ideas and skills over longer periods of time.

There is an Answer Key at the end of this part of the book. Check the Answer Key after every scenario to ensure that you are accurately practicing what you have learned. If you struggle to complete any scenarios, try to find someone who can guide you through them.

57

CONTENTS

Section 1
REVIEW

1. Solve each equation, without a calculator.

    a. $3x+5x=32$ b. $-\dfrac{2}{3}y=4$ c. $-6z+27=3z$

2. Solve each equation, without a calculator.

    a. $7A-1.2=-3A-6.2$ b. $\dfrac{2+4B+15+5}{4}=3$ c. $0.4=3-\left(C-5\right)$

3. A toy company sends researchers to a mall to find out how many kids know about their most recent toy. Kids who are willing to take a quick survey are asked a simple yes or no question. The results are shown. After 5 groups are surveyed, the results show that 70% of all of the kids who took the survey answered the question with a response of "Yes." In the fifth group surveyed, what percent of the kids responded with a "No?"

Group Number	Number of Kids Surveyed	Number of "Yes" Responses
1	100	45
2	100	80
3	100	72
4	100	64
5	100	?

4. Use a calculator, as needed, to solve each equation.

    a. $0.32h=327.68$ b. $g+0.08g=9$ c. $35.2=f-0.12f$

Section 2
INTRODUCTION TO PERCENTS

5. Some percentages can be computed mentally, without doing calculations on paper or on a calculator.

    a. What is 50% of 60?

    b. How much money is 25% of $80?

    c. 20 people said they would help clean up trash along the roads on Saturday morning, but only 80% of them actually came to help. How many people did not show up?

Other percent calculations, like 27% of 119, or 6% of 592, require more time and can be done more quickly with a calculator. This brings up an important point, though. A calculator is only helpful if you input correct values.

6. For example, if you try to determine 2% of 100 by making a calculator compute 2x100, the calculator will give you a result of _____. Explain why this result is not reasonable.

7. If you can estimate percentages before you do calculations, you might notice your errors if you make mistakes. Estimate each value below, rounded to the nearest whole number. Do not use a calculator.

    a. 98% of 15   b. 51% of 32   c. 23.5% of 60

8. What fractional amount of the total figure is shaded in each image below?

    a.   b.

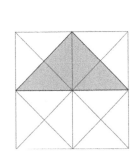 

9. Convert each of the previous fractions to percentages. If needed, use a calculator.

10. Write an expression that represents each of the following.

    a. 85% of *A*   b. 103% of *B*   c. 500% of *C*   d. 0.85% of *D*

Section 3
PERCENT CHANGES: INCREASES & DECREASES

63

11. If 40 increases by 50%, its new value is _____.

12. If 40 increases by 51%, what is its new value?

13. Fill in each blank below.

     a. If 100 increases by 35% its new value is _____.

     b. If 100 decreases by 2% its new value is _____.

     c. If 48 increases by 130% its new value is _____.

     d. If 48 decreases by 25% its new value is _____.

14. Write an expression that represents each of the following.

     a. E decreases by 4%

     b. F increases by 12%

     c. 250% more than G

     d. 80% less than H

15. Notice your expressions in the previous scenario. There are some common structures that you can identify in those expressions. Fill in each blank below.

     a. If K increases by 90% its new value is _____.

     b. If K decreases by 19%, its new value is _____.

     c. If K increases by 200% its new value is _____.

     d. If K decreases by 100%, its new value is _____.

16. Notice your expressions in the previous scenario. There are some common structures that you can identify in those expressions. Fill in each blank below.

     a. If K becomes $3.25K$, it has increased by _____%.

     b. If K becomes $0.77K$, it has decreased by _____%.

17. Compute the percent by which the initial value must change in order to become the final value.

    a. initial value: 100   final value: 125   percentage change: _____

    b. initial value: 100   final value: 75   percentage change: _____

    c. initial value: 100   final value: 2   percentage change: _____

18. Fill in the blank to show the percentage change.

    a. initial value: 100   final value: 200   percentage change: _____

    b. initial value: 100   final value: 400   percentage change: _____

    c. initial value: 100   final value: 582   percentage change: _____

19. Compare each expression shown below to an original amount of T. State whether T has increased or decreased and identify the percent by which T has changed.

    a. $0.41T$ b. $1.99T$ c. $3.12T$ d. $4T$

20. Answer each of the following questions.

    a. If a number doubles, by what percent has it increased?

    b. If a number quadruples (becomes 4 times larger), by what percent has it increased?

    ★c. If a number becomes one-tenth of its original value, by what percent has it decreased?

    ★d. If a number becomes one-third of its original value, by what percent has it decreased?

21. When you look at a digital screen, the image is actually a large number of colored dots (called pixels) packed very closely together. Over time, the number of pixels contained in each square inch of screen space has increased. For example, a digital screen from 2007 contains 163 pixels in every square inch of screen space. A digital screen from 2014 contains 401 pixels per square inch. By what percent did the number of pixels per square inch increase from 2007 to 2014?

22. Alyssa, Beth, and Chris, each win a $1,000 award and deposit their prize money into their savings accounts. Before the deposit, Alyssa's account balance is $500,000, Beth's account balance is $50,000, and Chris's account balance is $5,000. Who is likely to be most grateful for this award and for what reason do you make your decision?

23. After each person in the previous scenario deposits their prize money, by what percent did their savings account balances increase?

24. ★Between the years of 2011 and 2015, the price of one share of Sears stock decreased by $42. In 2015, it was worth $32 per share. During this time period, Sears began closing many of their store locations as well. In 2011, there were 4,010 Sears stores, but there were only 1,672 stores remaining in 2015. From 2011 to 2015, which quantity decreased by a greater percent, the stock price or the number of store locations?

25. In 2015, a study counted 1,531 complaints against police. During that same year, several groups of police officers in the US and the UK agreed to have recording cameras attached to their uniforms to research the impact of these cameras. During the 12 months of the study, there were only 113 complaints against police. After these results were published, an online journal wrote that the number of complaints dropped 98% over the course of the one-year study. Several websites restated this conclusion. Do you agree with this published conclusion?

26. The chart shows the population of a species of bird over a five-year period.

    a. By what percent did the population change from 2012 to 2013?

    b. Between which consecutive years did the population change by the smallest percent?

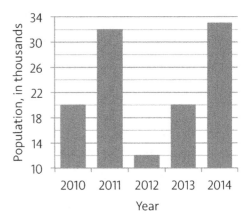

Section 4
WRITING EQUATIONS TO CALCULATE PERCENTS

27. Write each statement as an equation. Do not solve the equation.

   a. The sum of a number and 10 is 13.

   b. The difference of 20 and a number is 3.

   c. 4 more than twice a number is 16.

28. Write each statement as an equation. Do not solve the equation.

   a. 6 more than a number is 13.

   b. A number increased by 17 is 30.

   c. A number is 8 less than 14.

   d. 21 is 13 less than a number.

The previous two scenarios do not involve percentages, but they are intended to help you think about writing mathematical statements as equations. The next scenarios will focus only on percentages.

29. Write each statement as an equation. Do not solve the equation.

   a. A number is 3% of 20.

   b. 205% of a number is 615.

   c. 30 is some percentage of 500.

30. Find the value of each number described in the previous scenario. Use a calculator if needed.

31. Consider the following question: What number is 8% of 17? Convert this statement into an equation.

32. Read the question and convert it into an equation. Do not solve these equations.

    a. 9% of 45 is what number?   b. What number is 110% of 30?

33. Solve each of the equations in the previous scenario.

34. Read the question and convert it into an equation. Do not solve these equations.

    a. 9% of what number is 18?   b. 30 is 120% of what number?

35. Solve each of the equations in the previous scenario.

36. Read the question and convert it into an equation. Solve the equation to answer the question.

    a. 320 is what percent of 200?   b. What percent of 80 is 52?

37. Read the question and convert it into an equation. Solve the equation to answer the question.

    a. 5 is what percent of 6?   b. What percent of 480 is 33?

38. Each of the following scenarios shows a percentage relationship that can be expressed as follows: ___% of ___ is ___. Read each scenario and write an equation that expresses the percentage relationship in that scenario. Do not solve the equations.

     a. At a football game, there are 52,000 people in attendance, but the stadium is only filled to 65% of its seating capacity. What is the seating capacity of the stadium?

     b. On average, 76.5% of the people who take the bar exam every year to become a licensed lawyer receive a passing score on the exam. If 55,200 people passed the exam this year, how many people took the exam?

39. Answer each of the questions in the previous scenario.

40. Each of the following scenarios is slightly more complex than the previous group of scenarios. These scenarios involve a quantity that changes by some percent to take on a new value. Read each scenario and write an equation that expresses the percentage relationship in that scenario. Do not solve the equations.

     a. In 2014, there were 492 billionaires living in the United States, which was 11.3% more than the year before. How many billionaires lived in the U.S. in 2013?

     b. After the 8% tax was calculated, the total cost of the meal was $47.52. What was the cost of the meal before the tax?

41. Answer each of the questions in the previous scenario.

42. At the Rodale Apple Festival last year, you could buy a bushel of apples for $30. Since then, there has been a severe drought. As a result, the price for one bushel increased this year by 18%. After this increase, what is the price for two bushels of apples?

43. If the value of a motorcycle in March of 2013 was D dollars, but the value decreased 15% over the course of the next 12 months, what is the motorcycle's value in March of 2014, written in terms of D?

44. On September 20th, the initial price of one share of a stock was $21.84. That price was 4% lower than the initial price on the previous day. What was the price of one share of that stock on the 19th?

45. The population of a species of bird was 20,000 in 2010. If the bird population decreased 35% from 2009 to 2010, what was the bird population in 2009?

Section 5
VARIOUS PERCENT SCENARIOS

46. Nadya's puppy weighed 8 pounds on June 10th. As her puppy grew, it's weight increased by 10% each day. How much did the puppy weigh on June 14th?

47. In 1980, a typical hard drive could hold 5 MB of data. By 1990, a typical hard drive had a storage capacity of 40 MB. By what percent did the storage capacity increase from 1980 to 1990?

48. The median price of a home in the U.S. peaked in April 2008 at the beginning of the financial crisis. Over the next four years, the median home price dropped 22% to $152,000 in April 2012. What was the median price of a home in April 2008?

49. ★In 2014, Netflix had a total of 8,103 programs available for viewing and 6,494 of these programs were movies while the remaining programs were TV shows. In 2016, the total number of available programs had decreased to 5,532 and 1,197 of these programs were TV shows, while the remaining programs were movies. From 2014 to 2016, did the number of movies or TV shows decrease by a greater percent?

50. At the store, you see that a hat costs $20. When you go to pay for it, you hand over a $20 bill but you are told that this is not enough. The final price of the hat is $21.60, after sales tax is added in. What is the sales tax rate, expressed as a percentage?

51. After you buy a bowl of frozen yogurt, you are asked to pay $4.88. If the sales tax is 8%, what was the cost of the frozen yogurt before the tax was calculated? How much did you pay in taxes?

52. In 2008, the endowment of a school was approximately $69,000,000. After the stock market fell in 2009, the endowment fell to $53,000,000. By 2011, the endowment had risen once again to the 2008 mark of $69,000,000.

  a. Calculate the percent change in the endowment from 2008 to 2009.

  b. Calculate the percent change in the endowment from 2009 to 2011.

  c. Explain why the percentages are different in parts a. and b.

53. In order to wrestle in a different weight class, Cody changed his diet and workout routine and gradually lowered his weight by 12% to 110 pounds. What was his original weight?

54. ★A new phone has a rectangular screen with a height that is 292 pixels greater than the width. If the perimeter of the phone's screen is 2,084 pixels, what are the dimensions of this screen, measured in pixels?

55. Another phone has screen dimensions of 320 pixels by 480 pixels. By what percent is the area of this phone screen larger than the area of the phone screen in the previous scenario?

56. Due to inflation, prices tend to rise every year, which means that one dollar will buy more today than it will in 10 years. For example, inflation caused prices to increase by 204% from 1980 to 2014. If a pair of shoes cost $100 in 2014, how much would you have paid for those shoes in 1980?

57. The value of a soccer team increases by 25% every year for 4 years in a row. By what percent did the value of the soccer team increase over the entire course of that 4-year period?

58. A recent study examined which eye color was most common among students at a particular school. The study found that 45% of the students have green eyes. Why is this an inaccurate conclusion?

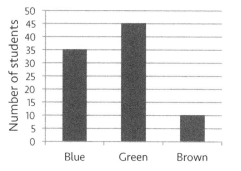

59. ★The Shake Shack sells a Caramel Classic milkshake and it has been really popular lately so they raised the original price by 10% to increase their profits. Unfortunately, the interest in that milkshake flavor went down dramatically. As a result, they lowered the price by 10%, which brought the price of the Caramel Classic down to $4.00. What was the original price of the Caramel Classic milkshake?

60. Many restaurants try to use 30% of the money they earn each week to pay their employees.

   a. If a restaurant earns an average of $21,000 in sales each week, how many employees can they hire if they pay their employees an average of $580 per week.

   b. Estimate the amount of money that a restaurant makes in sales each week if they have 15 employees and they pay each of them an average of $650 each week?

61. The chart shows a family's water bill for the month of September. The total bill is composed of water charges and extra fees.

   a. What percent of the total bill is composed of the extra fees?

   b. If the water bill is paid after the current due date, the amount due will increase by _____ percent.

   c. What was the percent change in the water bill from August to September?

Billing Detail	
Amount owed from last bill $85.40	
Total payments received 85.40	
Remaining balance 0.00	
Current water charges 67.95	
Extra fees 7.55	
Amount due ON or BEFORE 10/14/16 ....$75.50	
Amount due AFTER the current due date $78.52	

62. In the previous scenario, the water bill for the month of August was 8% higher than the water bill for the month of July. What was the water bill for the month of July?

63. If the value of an investment increases by 10% every year, by what percent will the value of that investment increase over a 7-year period?

64. If the value of a $10,000 investment decreases by 50% every year, how many years will it take for the investment to lose all of its value (to have a value of $0)?

65. ★Three competing mattress companies try to get you to buy from them by offering discounts on your purchase.

    Matt's Mattresses will give you both a 10% and a 20% discount.

    Betty's Beds offers 2 discounts as well, a 5% and a 25% discount.

    Not to be outdone, Sleepy Sal's will give you a 14% and a 16% discount.

Each company will let you use their discounts in either order that you choose. What is the best overall percentage discount that you can receive?

66. The values below show the increase in costs in the NFL from 1998 to 2014. Each value is an average.

Cost to attend a game	Salary of NFL player	Value of an NFL team
1998 – $55	1998 – $1.4 million	1998 – $300 million
2014 – $123	2014 – $2.1 million	2014 – $1.4 billion

    a. Which of the 3 categories has increased by the greatest percent?

    b. Calculate the average rate of increase from 1998 to 2014, in dollars per year, for each of the three categories.

67. In Maine, lobster fishing is an important business. To help preserve the lobster population, about 85% of the lobsters that are caught are thrown back into the ocean. In 2016, about 130 million pounds of lobster were caught and kept to be sold. Calculate how many pounds of lobster were caught in 2016.

68. ★You spend 50% of the money in your savings account every year for 2 years in a row. By what percent does the amount of money in your savings account decrease during this 2-year period?

69. ★A local shoe store is closing down so they offer a one-day only 50% off sale. You are one of the lucky shoppers who also receive a coupon for 10% off of your total discounted purchase. With this coupon, what percent of the original price of a pair of shoes will you end of paying?

70. Each year, the number of people in a city increases by 10%. After 5 years, will the city's population have increased by more than, less than, or exactly 50%? How do you know this?

Section 6
CUMULATIVE REVIEW: PART 1

71. Simplify each expression as much as you can.

    a. $3-(x-1)$
                                  b. $100-(-5)^2$

72. Simplify the expression as much as you can.

$$3-4\left[2-(-1)^3\right]$$

73. What is the largest number that is a factor of both 48 and 60?

74. What is the greatest common factor of 60 and 80?

75. If you write down the multiples of 6 and the multiples of 15, what is the smallest number that will be in both of those lists?

76. What is the least common multiple of 8 and 20?

77. Consider the equation shown.

$$f=11-6m$$

    a. What is the value of f when m is replaced with 5?

    b. What is the value of m when $f=2$?

78. You form a box by taking rectangular pieces of cardboard and taping them together. When the box is formed, each edge of the box is taped from end to end. The box is shown.

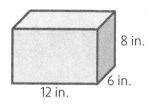

8 in.

6 in.

12 in.

    a. How many pieces of cardboard did you tape together to make the box?

    b. How many inches of tape did you unroll to tape all of the edges of the box?

    c. How many square inches of cardboard did you put together to make the box?

79. When the box in the previous scenario is empty, how much air does the box contain inside it, measured in cubic inches?

80. Consider the equation shown.

$$6g - 20h = 12$$

    a. What is the value of g when $h = 0$?

    b. What is the value of h when $g = 0$?

Section 7
INTRODUCTION TO RATES

81. A worm moves 5 centimeters per minute. How far will it move in 1 hour?

82. If a family spent $9,360 on food in one year, how much did they spend each month, on average?

83. Suppose the hair on your head grows about 0.5 inches per month.

    a. After 6 months, by how many inches will the length of your hair increase?

    b. If you never cut your hair, how many inches will your hair grow in *Y* years?

84. Sound travels about 20,400 meters per minute.

    a. What is the distance that sound travels every second?

    b. How far does sound travel every hour?

85. Sound travels faster through water than it does through air. It travels through air at 340 meters per second, and it travels through water at 1,500 meters per second. Suppose an explosion is detonated on the surface of the water far out in the ocean. A ship is positioned 5,000 meters away. If a diver is under the ship and a sailor is on the deck of the ship, how much sooner does the sound reach the diver than it reaches the sailor?

86. Penny purchased 15 gallons of gas for $34.50. Beth bought 20 gallons for $46.40.

    a. Who got a better deal buying her gas?

    b. How much would each person pay for 10 gallons of gas?

87. ★The international space station (ISS) travels at a rate of approximately 17,000 miles per hour, which causes it to complete one orbit around the entire earth every 90 minutes. How many sunrises will an astronaut on the ISS see after...

    a. 4 days in orbit?   b. *D* days in orbit?

88. Manny and Julian are out on a lake in a large but old fishing boat when it begins leaking. By the time they find a way to plug the leak, the boat has taken in 120 gallons of water. At 2:00pm, Manny grabs a bucket and begins tossing the water out of the boat at a rate of 3.5 gallons per minute. Manny has been removing water for 4 minutes before Julian finds another bucket and joins him, removing water at a rate of 2.5 gallons per minute. At what time will the boat contain only 10 gallons of water?

89. A car is moving along the highway. At 11:00am, it passes mile marker 60. At 11:08am, it passes mile marker 66. How fast is the car moving, in miles per hour?

90. If you mail a large envelope, the first ounce will cost $0.94, while each additional ounce will be priced at a different rate. Based on the information in the table, what is the rate for each additional ounce? Identify the numerical value of the rate and express the rate using proper units.

Weight (ounces)	3	6	8	14
Price (dollars)	1.36	1.99	2.41	3.67

91. The table shows how many liters of water would be purchased for a wedding, based on the number of guests attending the wedding. Assume the amount of water purchased increases at a constant rate as the number of guests increases.

Number of guests	120	200	320
Liters of water purchased	35	55	85

   a. How many liters of water would be purchased at a wedding with 322 guests?

   ★b. If 72 liters of water are purchased for a wedding, estimate the number of guests that will attend the wedding.

   ★c. If 0 guests attend a wedding, does the pattern in the table suggest that 0 liters of water would be purchased?

Section 8

USING GRAPHS TO CALCULATE RATES

92. During a winter blizzard, the amount of snow on the ground slowly increased each minute. The bar graph shows how the depth of the snow increased during a given time frame. The snow fell at a constant rate from 6:00am until 8:00am.

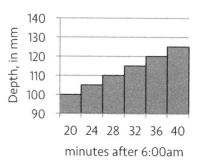

a. What was the rate at which the snow was falling, in millimeters per hour?

b. What was the depth of the snow at 7:00am?

★c. From 8:00am until 8:30am, the snowfall rate was 50 millimeters per hour. What was the depth of the snow at 8:30am?

93. When the water is boiling, the steam slowly evaporates into the surrounding air. The following graph show how quickly the amount of water in a pot decreases as time passes. What is the rate shown in this graph? Identify the numerical value of the rate and express the rate using proper units.

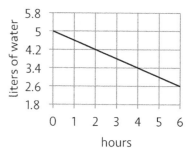

94. Draw a graph to represent each rate described below. Label each axis with numbers and titles to show that the numbers on each axis represent a specific type of measurement.

a. A worm crawls 10 inches per minute.

b. Lemonade costs 50 cents per cup.

95. After school ends one Friday, you get in the car and travel to another state to spend the weekend away from home. The graph below displays information from a portion of your trip.

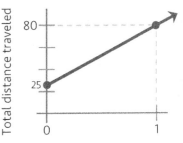

Total distance traveled

Hours after 4:00pm

   a. If the car maintains a constant speed, how fast is the car traveling, in miles per hour?

   b. How far away from school are you at 4:00pm?

   c. Write an expression for the total distance that you have traveled after *H* hours, where *H* is the number of hours after 4:00pm.

   d. Using your expression in part c., at what time you will be 201 miles away from school?

96. The graph displays the population of the state of California from 1940 to 2000.

   a. During which time period did the population increase by a greater percent, 1950–1970 or 1970–1990?

   b. Estimate the rate at which the population increased each year from 1950 to 1990.

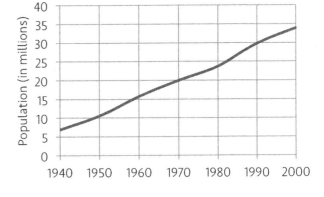

97. The graph displays how the boiling point of water changes as the elevation changes.

   a. At what rate does the boiling point decrease, in degrees per 1000ft?

   b. What is the boiling point of water at the top of Mount Everest, which has an elevation of 29,000ft?

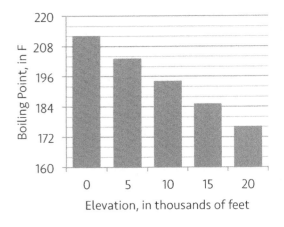

Section 9
RATES IN EQUATIONS

98. Consider the equation $y = 8x$.

   a. What is y if $x = 10$?

   b. What is the x when $y = 48$?

99. Consider the equation $T = 45f + 70$.

   a. Find the value of T when f is 2.

   b. What is the value of f when $T = -20$?

100. The equation $P = 800 - 23y$ shows how the bison population, P, changed every year from 1900 to 1910, where y is the number of years after 1900.

   a. Did the population increase or decrease from 1900 to 1910?

   b. What was the population in 1902?

   c. What was the annual rate of change in the bison population each year from 1900 to 1910?

   d. What does the 800 represent in the equation?

101. A wood burning stove heats a cabin in the winter, but it makes the air very dry. A pot of water is placed on top of the stove to add moisture to the air as the water evaporates. The amount of water in the pot is given by the equation $W = 5 - 0.25h$, where W is the amount of water in the pot, in liters, after the pot has been sitting on the stove for h hours.

   a. How much water is in the pot after it has been sitting on the stove for 6 hours?

   b. In this equation, what do the numbers 5 and 0.25 represent?

102. In the previous scenario, how many hours will it take for all of the water in the pot to evaporate?

103. A toll road charges an initial amount of $0.25 for driving on the toll road plus 8 cents per mile that you travel.

    a. How much will you have to pay for your toll if you drive on the toll road for 40 miles?

    b. How many miles can you drive on this road if you want to keep your toll below $8?

104. A skydiver typically starts a jump from an altitude of around 12,500 feet. If you ever decide to go skydiving, when you jump out of the airplane, your falling speed will get faster (accelerate) until you reach a terminal velocity. When you reach terminal velocity, you will continue falling at a constant rate until you open your parachute. Once you begin falling at terminal velocity, your altitude can be represented by the equation $A = 9{,}000 - 200t$, until you deploy your parachute. In this equation, A is your altitude in feet and t is the number of seconds that have passed after you begin falling at terminal velocity.

    a. What is your altitude 10 seconds after you reach terminal velocity?

    b. How many seconds can you fall at terminal velocity if you need to deploy your parachute at an altitude of 2,000 feet?

    c. What is your altitude at the moment you begin falling at terminal velocity?

105. Graph the previous scenario to show how your altitude changes each second after you reach terminal velocity.

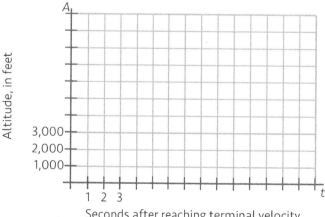

Seconds after reaching terminal velocity

Section 10
CUMULATIVE REVIEW: PART 2

106. Solve each equation.

    a. $0.25a + a = 25$ b. $3 + 3b = b - 5$

107. Solve each equation.

    a. $3 + \dfrac{1}{4}c = -5$ b. $\dfrac{12 + 7 + 15 + d}{4} = 12$

108. Which equation has the same solution as the equation $4x + 15 = 3 - 2x$?

    a. $2x - 12 = 0$ b. $6x + 15 = 3$

109. After learning how to program, you create a game and you keep track of the results. You make it possible to win the game to keep players interested. You make it hard to win, though, because if the game is too easy, users will not be interested in playing. Every time the game has been played 200 times, you track how many times the game was won and then you adjust the difficulty of the game to try to affect the winning percentage.

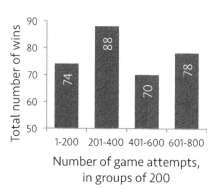

    a. What was the overall winning percentage during the first 800 times that the game was played?

    b. You set a goal to keep the overall winning percentage at 40%. In order to achieve your goal, you need the players to win _____ % of the next 200 attempts (801-1,000).

110. In 2015, Americans spent $350 million dollars on Halloween costumes for their pets. The rest of the money was spent on costumes for children and costumes for adults.

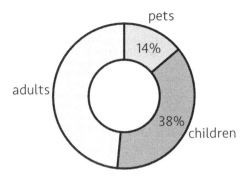

    a. How much money was spent on costumes in 2015?

    b. Costumes for adults accounted for what percent of the total spending?

    c. If the amount of spending decreased 11% from 2014 to 2015, how much money was spent on costumes in 2014?

111. The average temperature during the month of August was 80.6°F.

    a. If that was 4% higher than the average temperature during the month of July, what was the average temperature in July?

    b. What was the average temperature in September if it was 6% lower than the month of August?

112. ★After you finish brushing your teeth one night, you forget to shut it off completely, but you do not realize it until the next day. The faucet drips at a rate of 20 drops per minute. Assume that it takes 1,000 drips to fill one cup. By 12:00am (midnight), the dripping faucet has wasted 5 cups of water.

    a. What time did the faucet start dripping?

    b. If the faucet wastes a total of 1 gallon (16 cups) of water before you turn it off, what time did you realize that the faucet was dripping and turn it off?

113. ★For the previous scenario, write an equation that relates the total number of cups wasted, C, to the number of hours that have passed after midnight, h.

Section 11
ANSWER KEY

1.	a. $x = 4$ b. $y = -6$ c. $z = 3$
2.	a. $A = -0.5$ b. $B = -2.5$ c. $C = 7.6$
3.	11% said No (89% said Yes)
4.	a. $h = 1{,}024$ b. $g = 8\frac{1}{3}$ c. $f = 40$
5.	a. 30 b. \$20 c. 4 people
6.	2% of 100 must be less than 100, but 2x100 is 200, which is more than 100.
7.	a. 15 ($\approx$ 100% of 15) b. 16 ($\approx$ 50% of 32) c. 15 ($\approx$ 25% of 60, or 60 $\div$ 4)
8.	a. $\frac{4}{16} \to \frac{1}{4}$ b. $\frac{10}{16} \to \frac{5}{8}$
9.	a. $0.25 \to 25\%$ b. $0.625 \to 62.5\%$
10.	a. 0.85A b. 1.03B c. 5C d. 0.0085D
11.	a. $40 + 0.5(40) = 40 + 20 = 60$
12.	b. $40 + 0.51(40) = 40 + 20.4 = 60.4$
13.	a. 135 b. 98 c. 110.4 d. 36
14.	a. 0.96E b. 1.12F c. 3.5G d. 0.2H
15.	a. 1.9K b. 0.81K c. 3K d. 0
16.	a. 225% b. 23%
17.	a. increase 25% b. decrease 25% c. decrease 98%
18.	a. increase 100% b. increase 300% c. increase 482%
19.	a. decrease 59% b. increase 99% c. increase 212% d. increase 300%
20.	a. 100% b. 300% c. 90% d. $66\frac{2}{3}\%$
21.	$\frac{238}{163} \approx 1.46 \to$ approx. 146%
22.	It will likely be Chris, because the prize money will increase his current savings account balance by the greatest percent.
23.	Ally's account increases by 0.2%, Bea's by 2%, and Chris's by 20%.
24.	The number of stores decreased by a greater percent. stores: $\frac{4{,}010 - 1{,}672}{4{,}010} = 58.3\%$ stock: $\frac{74 - 32}{74} = 56.8\%$

25.	The percent calculation is incorrect. There was actually a 93% decrease.
26.	a. $66\frac{2}{3}\%$ b. 2010-2011: 60%
27.	a. $n + 10 = 13$ b. $20 - n = 3$ c. $2n + 4 = 16$
28.	a. $n + 6 = 13$ b. $n + 17 = 30$ c. $n = 14 - 8$ d. $21 = n - 13$
29.	a. $n = 0.03(20)$ b. $2.05n = 615$ c. $30 = \frac{n}{100}(500)$ or $30 = n(500) \to$ if n is the decimal form of the percentage.
30.	a. 0.6 b. 300 c. 6%
31.	$n = 0.08(17)$
32.	a. $0.09(45) = n$ b. $n = 1.10(30)$
33.	a. $n = 4.05$ b. $n = 33$
34.	a. $0.09n = 18$ b. $30 = 1.20n$
35.	a. $n = 200$ b. $n = 25$
36.	a. $320 = \frac{x}{100}(200) \to 320 = 2x \to x = 160\%$ b. $\frac{x}{100}(80) = 52 \to \frac{4}{5}x = 52 \to x = 65\%$
37.	a. $5 = \frac{x}{100}(6) \to 5 = \frac{3}{50}x \to x = 83\frac{1}{3}\%$ b. $\frac{x}{100}(480) = 33 \to 4.8x = 33 \to x = 6.875\%$
38.	a. $52000 = 0.65x$ b. $55200 = 0.765x$
39.	a. 80,000 people b. approx. 72,157 people
40.	a. $1.113x = 492$ b. $1.08x = 47.52$
41.	a. 442 billionaires b. \$44
42.	$\$35.40 \times 2 = \$70.80$
43.	0.85D
44.	$0.96p = 21.84 \to \$22.75$
45.	approx. 30,770 birds solve: $0.65x = 20{,}000$
46.	About 11.7 pounds $8(1.1) = 8.8 \to 8.8(1.1) = 9.68$ $\to 9.68(1.1) = 10.65 \to 10.65(1.1) = 11.7$ More concise: $8(1.1)^4 \approx 11.7$

47.	$40-5=35 \rightarrow 35 \div 5=7 \rightarrow 700\%$
48.	solve $.78x=152000$; approx. $195000
49.	Movies decreased by 33.2%, which is a greater percent than TV shows (25.6%).
50.	$1.60 \div 20 = 08 \rightarrow 8\%$ sales tax
51.	cost: $\approx$$4.52 tax: $\approx$36 cents
52.	a. $\approx$23.2% b. $\approx$30.2% c. The amount of change of $16,000,000 is a larger percent of $53,000,000.
53.	125 lbs.
54.	a. Let width = w, height = h and $h=292+w$. Solve: $2w+2(292+w)=2{,}084 \rightarrow w=375$ Dimensions: 375 by 667 pixels
55.	approx. 62.8% (the screen increased from 153,600 pixels to 250,125 pixels)
56.	Around $33 $\rightarrow$ solve: $x+2.04x=100$
57.	Approx. 144% $\rightarrow 1.25 \cdot 1.25 \cdot 1.25 \cdot 1.25 \approx 2.44$
58.	It is 45 out of 90, which is 50%
59.	$4.04
60.	a. 10.86 → approx. 11 employees b. $32,500
61.	a. 10% b. 4% c. it decreased about 11.6%
62.	$\approx$$79.07
63.	Approx. 94.9%
64.	It never will.
65.	Betty's = 28.75% overall Matt's = 28% Sal's = 27.76%
66.	a. Value of an NFL team ($\approx$367%) b. game: $\approx$$4.25/yr, salary: $43,750/yr team value: $\approx$$68.75 million/yr
67.	About $866\frac{2}{3}$ million pounds. 130 million is 15% of the total amount caught. Solve the equation $.15x = 130$.
68.	75%
69.	90% of 50%, or 45% of the original price
70.	More than 50%. After each 10% increase, there are more people in the city, so each 10% after the first is increase of more people than the previous increase.
71.	a. 4 – x b. 75
72.	–9
73.	12
74.	20
75.	30
76.	40
77.	a. $f=11-6(5) \rightarrow f=11-30 \rightarrow f=-19$ b. $2=11-6m \rightarrow -9=-6m \rightarrow m=1.5$ or $\frac{3}{2}$

78.	a. 6 b. 4(12) + 4(6) + 4(8) = 104 inches c. 2(12x8)+2(6x8)+2(12x6) = 432 in^2
79.	12x8x6 = 576 in^3
80.	a. $6g-20(0)=12 \rightarrow 6g-0=12 \rightarrow g=2$ b. $6(0)-20h=12 \rightarrow -20h=12 \rightarrow h=-\frac{12}{20}$ $\rightarrow h=-\frac{3}{5}$ or -0.6
81.	$5 \times 60 \rightarrow 300$ centimeters
82.	$9360 \div 12 \rightarrow$ $780 per month
83.	a. 3 inches b. 6Y
84.	a. 340 meters b. 1,224,000 meters
85.	About 11.4 seconds sooner (3.3 vs. 14.7 sec)
86.	a. Penny ($2.30/gallon vs. $2.32/gallon) b. Penny: $23.00; Beth: $23.20
87.	a. 64 b. 16D
88.	2:20 pm
89.	45 miles per hour (6 miles in 8 minutes)
90.	$0.21 per ounce
91.	a. 85.5 liters b. 268 people c. No, there would be 5 liters purchased
92.	a. 75 mm per hour (1.25 mm per minute) b. 150 mm c. 250mm
93.	Decrease 0.4 liters per hour
94.	Possible graphs are shown below: a. b.
95.	a. 55 mph b. 25 miles c. 25 + 55H d. solve: 25 + 55H = 201; H=3.2 → 7:12pm
96.	a. 1950-1970 (100%) is greater than 1970-1990 (50%) b. increased $\approx$20 million over 40 years, which is a rate of 500,000 people per year
97.	a. 1.8° per 1000 ft b. 159.8°
98.	a. $y=8(10) \rightarrow y=80$ b. $48=8x \rightarrow x=48 \div 8 \rightarrow x=6$
99.	a. $T=45(2)+70 \rightarrow T=160$ b. $-20=45f+70 \rightarrow -90=45f \rightarrow -2=f$
100.	a. decrease b. $800-23(2) \rightarrow 754$ bison c. 23 bison/year d. the population in 1900
101.	a. 3.5 liters b. 5 is the original amount of water in the pot. 0.25 is the rate at which the water evaporates, in liters per hour
102.	20 hours
103.	a. 0.25 + 0.08(40) = $3.45

	b. solve: 0.25 + 0.08m < 8; 96 miles or less
104.	a. $A = 9{,}000 - 200(10) \rightarrow A = 7{,}000$ ft b. $2{,}000 = 9{,}000 - 200t \rightarrow -7{,}000 = -200t$ $\rightarrow t = 35$ sec c. 9,000 ft
105.	
106.	a. $a = 20$ b. $b = -4$
107.	a. $c = -32$ b. $d = 14$

108.	Equation b.
109.	a. 38.75% (310 wins out of 800) b. 45% (90 wins out of 200)
110.	a. \$2.5 billion b. 38% c. ≈\$2.81 billion
111.	a. 77.5°F (solve 1.04T = 80.6) b. ≈75.8°F (compute 0.94x80.6)
112.	a. 7:50pm (4h 10min before midnight) b. 9:10am
113.	$C = 5 + 1.2h$